Picture Credits:
National Aeronautics and Space Administration (NASA):
8 (top left), 13 (right), 31 (top right), 33 (bottom right), 35 (right),
36 (centre), 37, 38l, 40l, 43 (bottom right)

# SPACE

## CONTENTS

# The Universe

The Universe is made up of vast empty regions, or space, along with billions and billions of galaxies and stars. All the galaxies are strung together, making the Universe look like a giant spider's web. The Sun, Earth, Moon and all the other planets belong to the Universe.

## How big is it?

Until the early 20th century, it was widely believed that the Universe consisted of only one galaxy – our Milky Way. Later, astronomers discovered that what they thought to be nebulae were, in fact, galaxies. It is estimated that there are about 100 billion galaxies in the Universe. This number is, however, increasing since new galaxies are still being formed.

▶ *The Universe may go on expanding forever...*

▶ *The Universe may grow infinitely, or it may reduce back to its original point*

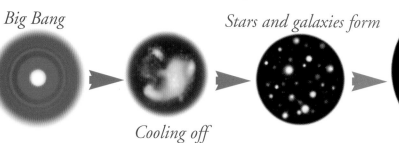

*Big Bang*

*Cooling off*

*Stars and galaxies form*

*The Universe today*

## Expanding Universe

As the Universe expanded, it started to cool down. Soon, tiny particles called atoms began to form. These atoms joined together to form stars and galaxies. Many scientists believe that the Universe is expanding even today. This theory was first put forth by the American astronomer Edwin Hubble. In the 1920s, Hubble noted that distant galaxies seemed to be moving away from our own. He also said that the greater the distance of the galaxies, the faster they were moving away.

*Scientists believe that the Universe was formed about 15 billion years ago. According to the Big Bang theory, all matter in the Universe was packed into a tiny fireball. This fireball gradually swelled up and started cooling down. The pressure inside the tiny point forced it to expand at a very fast rate, thus giving form to the Universe*

## Growing or shrinking?

So, will the Universe stop expanding sometime in the future? Or will the galaxies just continue to move away from each other? There are various theories regarding the future of the Universe. According to the 'open universe' theory, the Universe will continue to expand forever. The 'closed universe' theory, on the other hand, suggests that one day the Universe will stop growing and start to reduce in size, until all matter is contained in a small point – like it was in the very beginning. At this point, therefore, the Universe could end or even be reborn!

*...or it can start contracting, until all matter is contained within a single point!*

▶ *The spectrometer helps produce a spectrum and measure some properties of light such as wavelength, energy and intensity*

# Galaxies

The Universe is made up of billions of galaxies. But what are galaxies? Galaxies consist of dust, gas and millions of stars that are held together by their own gravity.

Spiral galaxy

Elliptical galaxy

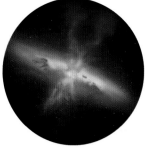
Irregular galaxy

## Types of galaxies

Galaxies are of various shapes and sizes. Depending upon the shape, galaxies are usually classified as spiral, elliptical or irregular. Spiral galaxies are shaped like a disk and have long spiralling arms in which new stars are born. Elliptical galaxies, on the other hand, are ball-shaped and consist of older stars. Galaxies that have no fixed shape are called irregular galaxies.

## Bright neighbours

As you already know, the Milky Way is just one of several galaxies in the Universe. Most of them are so far away from us that we cannot see them. However, a few galaxies have been found to lie close to our own. The Andromeda Galaxy, also called Messier Object 31 (or M31), is one such galaxy. It is almost twice the size of the Milky Way.

# Collision course

Do you know that galaxies have been known to collide into each other? This happens when two galaxies come near each other due to the pull of gravity. In the course of the collision, the galaxies can eventually merge and form a single galaxy. Since the stars in the colliding galaxies are usually too far apart, the merging of two galaxies does not create dangerous explosions.

## FACT FILE

- **Diameter of the Milky Way**
  About 100,000 light-years
- **Diameter of the Andromeda**
  About 200,000 light-years
- **Distance between Andromeda and Earth**
  About 2.2 million light-years
- **Distance between the Sun and the centre of the Milky Way**
  About 27,000 light-years
- **Average thickness of the Milky Way**
  10,000 light-years
- **Thickness of the Milky Way at the centre**
  About 30,000 light-years

*In about four billion years, the galaxy Andromeda could crash into the Milky Way. The Andromeda is believed to be speeding towards us at nearly 300 km per second. The two galaxies are expected to merge to form a new elliptical galaxy!*

*The Milky Way is a spiral galaxy and consists of over 200 billion stars as well as the Solar System. The Solar System is located on one of its arms, the Orion Arm*

# Starry Skies

**Looking up at the star-studded night sky it is difficult to believe that stars are, in fact, huge balls of hot gas and dust. New stars are born in clouds of dust and gas called nebulae. A star can live for millions, or even billions, of years before dying out.**

## Giants of the sky

Stars have been categorised into giant and dwarf stars depending upon their size. Supergiant stars are the biggest in the Universe. These stars are over 400 times larger than even the Sun. Giant stars are slightly smaller than supergiants, while the smallest stars are the red dwarfs. Our Sun is a medium-sized, yellow dwarf star.

## Clusters above

Most stars are found in groups. The Sun, a single star, is one of the rare cases. Many stars form pairs and are called binary stars. They orbit around a common centre of gravity. Stars can also belong to larger groups called clusters. Astronomers have broadly classified star clusters into open and globular clusters.

▲ *Globular clusters are formed when hundreds of thousands of stars come together in a tight ball. These may consist of stars as old as 12 billion years!*

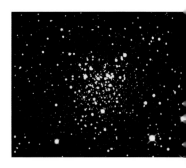

▲ *Open clusters are a loo group of about a dozer to hundreds of stars*

# Pictures in the sky

If you look up at the stars, you might be able to form patterns and outlines of animals and other shapes. In fact, ancient people identified groups of stars such as the hunter, bear, crab, dragon, and many other kinds of figures. Modern astronomers call these star pictures 'constellations'.

*Famous constellations include The Great Bear, The Little Bear and the Orion, commonly known as The Hunter. Orion is usually identified by a row of three bright stars that form a belt near the middle*

*Astronomers recognise a total of 88 constellations including the 12 signs of the zodiac. In ancient times, sailors used to find their way through the oceans by looking up at the constellations*

# Death of a Star

Stars do not live forever. Like the Universe, stars also expand as they grow older. Smaller stars, such as our Sun, swell up and eventually fade out. Massive stars, on the other hand, glow brighter as the end nears, before blowing apart in a huge explosion.

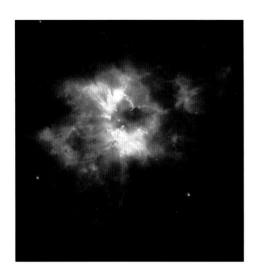

## A star is born

Stars are formed when gravity forces hydrogen gas to form a dense, spinning cloud. As the cloud spins faster, hydrogen atoms begin to bump into each other, releasing a great amount of heat that causes the gas to glow. Over time this glowing cloud of gas, called protostar, continues to grow and becomes a star.

▲ *Red giant stars formed by medium-sized stars could continue to lose their brightness to become a white dwarf, before finally fading away*

## Supernova

As the star continues to glow, the outer layer of gases begins to expand and cool. Meanwhile the hydrogen in the inner core is converted into helium, making the core contract. The cool outer layer starts to get a red glow. At this stage the star is called a red giant.

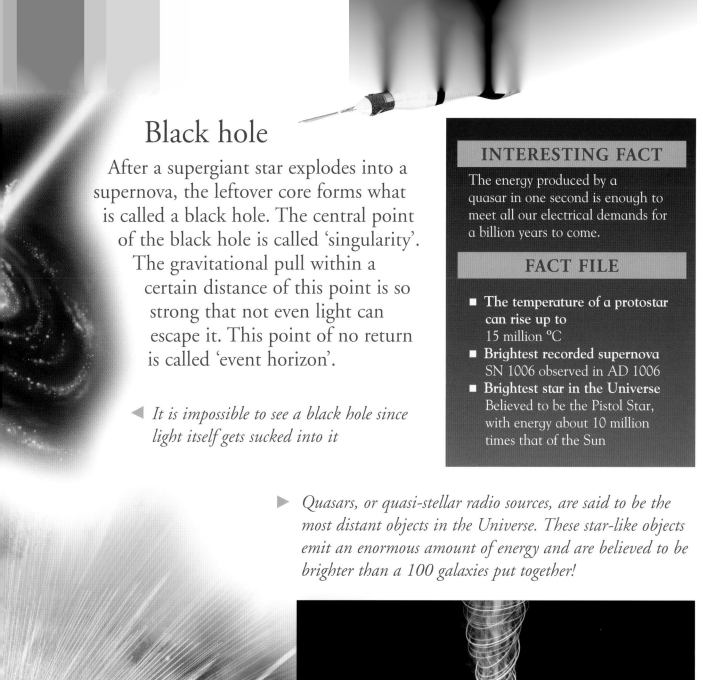

# Black hole

After a supergiant star explodes into a supernova, the leftover core forms what is called a black hole. The central point of the black hole is called 'singularity'. The gravitational pull within a certain distance of this point is so strong that not even light can escape it. This point of no return is called 'event horizon'.

◄ *It is impossible to see a black hole since light itself gets sucked into it*

► *Quasars, or quasi-stellar radio sources, are said to be the most distant objects in the Universe. These star-like objects emit an enormous amount of energy and are believed to be brighter than a 100 galaxies put together!*

◄ *Massive stars that form red giants die in a huge explosion called the supernova. As it explodes, the star often shines a billion times brighter than our Sun*

# The Solar System

**The Sun, the nine planets and their moons, asteroids, comets and meteoroids are together known as the Solar System. Our solar system is located on the edge of one of the spiral arms of the Milky Way.**

## Know your Solar System

The Solar System is shaped like an egg.
Until over 4 billion years ago, the Solar System was
a huge mass of gas, rocks and icy particles that drifted about in the Milky Way.
This mass apparently began to squeeze together, creating heat in the process.
Finally the centre exploded to create the Sun.
The nine planets were formed soon after.

Uranus

Jupiter

Saturn

Mars

Earth

Venus

Mercury

## Heavenly bodies

A planet is a large heavenly object that revolves around a star. There are nine known planets in our Solar System. The planets closest to the Sun are known as the inner planets. These planets are largely made up of rocks and include Mercury, Venus, Earth and Mars. Jupiter, Saturn, Uranus and Neptune are called the outer planets.

# The giant

Jupiter, the fifth planet from the Sun, is the largest planet in the Solar System. It also rotates faster than any other planet and has the most number of moons. An amazing fact about Jupiter is that a huge thunderstorm is believed to have been raging on its surface for more than 300 years. The region of this storm is called the Great Red Spot.

*Astronomers believe the Solar System was born about 4.5 billion years ago*

Pluto

Neptune

*Outer planets are composed of gases and, hence, also known as 'gas giants'. Originally only eight planets were known to us. The ninth planet, Pluto, was not known to exist until 1930*

*Mercury, which is the closest planet to the Sun, is about as big as our Moon and is the hottest planet in the Solar System, while Pluto is the smallest as well as the coldest planet. Mercury takes only 88 days to complete one orbit. The farthest planet, Pluto, takes over 248 years to go around the Sun*

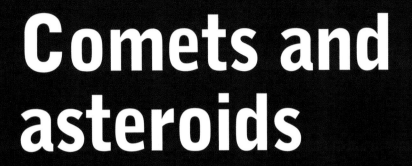

# Comets and asteroids

Apart from the Sun and the nine planets and their moons, our Solar System contains several pieces of rocks, metal and ice. These objects are asteroids, meteors and comets. Like planets, these objects also orbit the Sun.

## Almost a planet

Asteroids are planet-like objects that go around the Sun. These rock pieces are smaller than planets and are found in various sizes – some can be huge with a diameter of about 1,000 kilometres.

▼ *There are several thousands of asteroids in our Solar System, and most of these are found in a region between Mars and Jupiter. This region is known as the Asteroid Belt*

*The fragments of a large meteoroid, called meteorites, may strike the Earth's surface causing big craters*

# Not a star!

Have you ever wished on a shooting star? You might be surprised to know that shooting stars are not really stars, but tiny pieces of burning rocks (called meteors)! These rock fragments are formed from meteoroids, which are leftover fragments of a comet or an asteroid. When a meteoroid enters the Earth's atmosphere, it may collide with air molecules in the atmosphere and burn up into a meteor.

# Comets

Comets are made up of ice and rock. When a comet passes close to the Sun, the outer layers of its icy core melts to release jets of gas and dust particles. These are pushed behind the comet to form the tail. The tails of some comets can be as long as 150 million kilometres!

◄ *Halley's Comet can be seen once every 76 years, which is the time it takes to complete its orbit around the Sun*

# The Yellow Dwarf

The flaming ball of fire that we call the Sun is, in fact, a medium-sized star. It is the Sun's gravitational force that holds the nine planets and all the smaller objects in the Solar System in their respective orbits. Compared to many other stars in the Universe, the Sun is smaller as well as much younger. It was formed about 4.5 billion years ago.

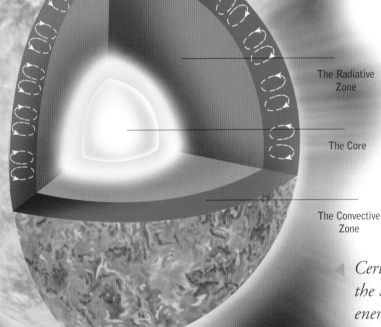

The Corona

The Chromoshere: a transparent layer above the photosphere

The Radiative Zone

The Core

The Convective Zone

## Hot to the core

Although it looks like a fireball, the Sun is not actually in flames. Have you seen how a piece of iron turns red when heated? The Sun's glow is similar to that piece of red-hot iron.

*Certain reactions at the centre of the Sun release a vast amount of energy, making it glow. This energy reaches us as heat and light*

# Flares and winds

Sometimes huge amounts of magnetic energy are discharged from the Sun and jets of gas are sent into space. These are called solar flares and cause a sudden increase in the brightness of the Sun. Solar flares can be followed by ejection of streams of electrically charged particles like protons and electrons. These are called solar winds.

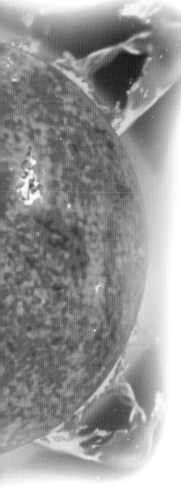

◀ *A solar flare is the sudden, intense eruption of high-energy radiation from the surface of the Sun*

## FACT FILE

- **Time the Sun takes to revolve around the Galaxy**
  An estimated 250 million years
- **Diameter of the Sun**
  About 1,392,000 million km
- **Gravity of the Sun**
  Nearly 28 times that of Earth
- **Temperature at the surface of the Sun**
  About 5,527°C
- **Temperature at the centre of the Sun**
  Over 15 million °C

# Hidden Sun

Can you imagine what it will be like if the Sun should suddenly disappear in the middle of the day! During a solar eclipse, the Moon passes in front of the Sun and blocks it from our view, either partially or totally. Solar eclipses can be total, partial or annular.

▶ *Collision between charged particles from the Sun and air molecules causes a luminous phenomenon called aurora. It is marked by streamers or bands of light*

# The Blue Planet

Earth is the fifth largest planet and located third from the Sun. It is the only planet with plenty of water and oxygen, both of which are essential for supporting life.

*▶ If you look down from space, the Earth would look blue because about 70 per cent of its surface is covered with water. Hence, it is also known as the 'blue planet'*

## Breath of life

Like all other planets, Earth too has an atmosphere, a film of gases that surrounds the planets. The Earth's atmosphere is divided into four layers – troposphere, stratosphere, mesosphere and thermosphere. Of all gases present in the Earth's atmosphere, nitrogen is most abundant. Minute traces of hydrogen, carbon dioxide and water vapour are also present.

Atmosphere

Outer Core

Inner Core

Crust

Mantel

*◀ Oxygen forms about 21 per cent of the total gases in the atmosphere. None of the other planets have an adequate quantity of oxygen, the gas that sustains life*

# Layers of Earth

Like its atmosphere, the Earth's surface is also divided into various layers. The innermost layer is called the core and is made up of iron and nickel. A part-rocky and part-molten layer called mantle surrounds the core. The temperature in the mantle is high enough to partially melt the rocks in this layer. The uppermost layer, called crust, is a thin layer of rocks. We live on the crust.

## Spinning Earth

The Earth, like all planets, revolves around the Sun. It also rotates on its own axis. It takes the Earth about 24 hours (or one day) to complete one rotation, while one revolution around the Sun is completed in approximately 365 days.

◄ *As the Earth rotates on its own axis, the side facing the Sun receives light while the side facing away is in darkness and experiences night*

# Many Moons

Just as the planets circle the Sun, there are smaller objects that move around the planets. These objects that revolve around the planets are called satellites, or moons. Some planets have more than one moon.

## Our Moon

The word 'moon' is used for the Earth's satellite, which is also called Luna. The Moon is made up of hard rocks. The surface is bumpy, with flat plains of powdered rock in some places and tall mountains in others. The surface also has huge holes called craters. There is no air or water on the Moon which makes it impossible to support life.

▼ *The Moon does not give off light of its own and only reflects the Sun's light*

## Lunar phases

The Moon does not always appear full and round. Its shape goes through various 'phases'. In the 'new moon' phase the Moon is between the Earth and the Sun. The side facing the Earth is away from the Sun and, hence, dark. This makes the Moon invisible from Earth. After a couple of days, a thin line lit by the Sun is seen. This is the crescent Moon. When the Moon is halfway around the Earth, the full sunlit side is visible. This is the 'full moon' phase.

# Other moons

Apart from the Earth, six other planets also have their own moons. Mars has two moons – Deimos and Phobos. It is thought that these moons may perhaps have been asteroids once. Saturn has over 30 moons, and until recently it was considered to have the largest number of moons. However, Jupiter is the new champion with over 60 moons found so far.

▶ *Among the more well-known moons of Jupiter are Ganymede, Io, Europa and Callisto. These moons were discovered by the famous Italian astronomer Galileo in 1610*

## FACT FILE

- **Distance of the Moon from Earth**
  About 384,400 km
- **Time taken by the Moon to go around the Earth**
  About 28 days
- **Size of the biggest crater on Earth's Moon**
  About 2,240 km in diameter
- **The largest of all planetary moons**
  Ganymede, about 5,262 km in diameter
- **Number of moons orbiting Uranus**
  27 recorded
- **Number of moons around Neptune**
  13 recorded
- **Number of moons orbiting Pluto**
  1, Charon

# The Red Planet

**Mars is the fourth planet from the Sun and has been a subject of interest for years. The surface of Mars has dry, desert-like regions containing rocks and soil rich in iron.**

## Life on Mars

Among all planets, Mars is the most similar to Earth. Since it is not too close to the Sun, Mars is never too hot. Winter temperature can, however, dip to as much as -130 degrees Celsius. Although the atmosphere of Mars is almost entirely made up of carbon dioxide, it is believed that it could support some forms of life that do not need oxygen to live.

## Frozen water

Scientists believe that water once flowed on Mars, probably billions of years ago. According to them, the planet then had lakes, rivers and perhaps oceans too. As Mars started to cool off, the water on its surface began to freeze. This could explain the presence of frozen water that exists permanently in ice caps at the poles of the planet.

▶ *Although the film of air around Mars is thinner than that of Earth, it is enough to support some type of life. That is why scientists believe that life could have once existed on Mars*

▼ *The presence of iron gives Mars a reddish tint. Hence, it is also called the 'red planet'*

# Evidence of life

For years, scientists have been trying to find out more about life on Mars. Several space explorations have been conducted in search of evidence for this. In 1984, scientists discovered a small meteorite in Antarctica. At 4.5 billion years old this meteorite was the oldest known rock. Scientists claimed that the meteorite contained substances similar to those found in certain tiny living organisms on Earth. However, there is still no clear proof as to whether life ever existed on Mars.

▼ *Mars is filled with craters, cliffs and volcanoes. One of its volcanoes, Mount Olympus Mons, is about three times the height of our own Mount Everest, and is also the largest known volcano in our Solar System*

# Lord of the Rings

Saturn is the second largest planet in the Solar System. It is almost 10 times bigger than the Earth, and has over 30 moons. Titan, its largest moon, is believed to have an atmosphere made up of nitrogen and hydrocarbon elements.

## Grounds of gas

Scientists believe that there is no solid surface on Saturn. Both Jupiter and Saturn are mostly made up of gases. Saturn has a rocky centre covered by a layer of liquid. This means that if we were ever to land on Saturn, we would be walking on air!

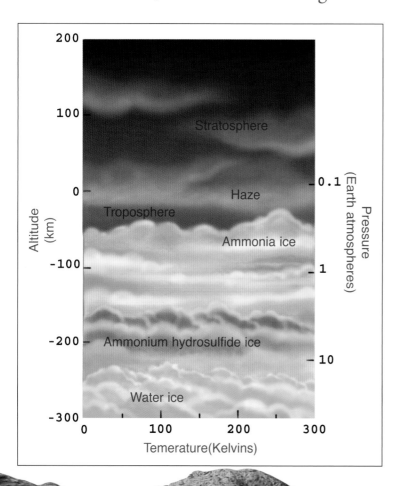

## Rings of ice

The sixth planet from the Sun, Saturn is most famous for its rings. It is encircled by seven main rings, which are made up of thousands of smaller rings. The rings are composed of dust particles and billions of ice pieces, with sizes varying from minute specks to mountain-like masses. These pieces of ice and dust keep moving around the planet.

◄ *The atmospheric layers on Saturn*

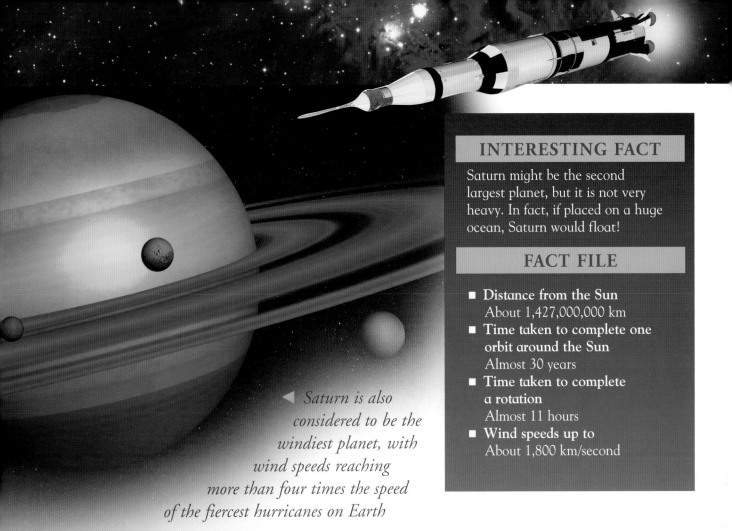

◀ *Saturn is also considered to be the windiest planet, with wind speeds reaching more than four times the speed of the fiercest hurricanes on Earth*

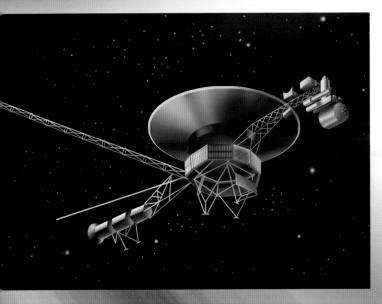

## Studying Saturn

In 1980 and 1981, two Voyager spacecraft visited Saturn. Voyager 1 and Voyager 2 took pictures of the planet and its moons. They helped us to learn more about Saturn's rings. They also discovered that one of its moons, Titan, has a thick atmosphere, mainly containing nitrogen, which is much like the Earth's atmosphere.

*Saturn's rings shine because the ice pieces reflect sunlight. All the four outer planets have rings, but only Saturn's are visible from Earth*

# Studying Space

**The study of the Universe and all its elements is called astronomy. For centuries humans have been intrigued by space. Ever since man first looked up at the sky, he has been trying to solve the mysteries of the Universe.**

## Ancient beliefs

In ancient times, people made up legends to explain matters they did not understand. According to one ancient Eastern legend, the Earth was a flat disc that rested on the backs of four elephants. These elephants stood on a tortoise, which, in turn, rested on a snake. Ancient Egyptians believed that the goddess of sky, Nut, arched over the Earth, with only her feet and fingers touching the ground. Her curved belly was said to form the star-spangled sky.

▼ *Ancient Babylonians were among the first known astronomers. They are believed to have recorded many of the stars and planets that we know of today, way back in 2000 BC. They even developed one of the earliest calendars based on astronomical events*

## First telescopes

The first telescope is believed to have been invented by a Dutch eyeglass maker called Hans Lippershey (1570-1619) in the year 1608. His telescope had two lenses with slightly different shapes, placed at the two ends of a tube. The well-known Italian scientist Galileo Galilei (1564-1642) made his own version of the telescope the following year. Using his telescope, Galileo first discovered Jupiter's moons.

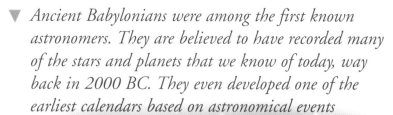

Many, many years ago, people believed that the Earth was the centre of the Universe and that the Sun, the Moon and the stars revolved around it. It was a Polish astronomer, Nicholas Copernicus (1473-1543), who established the theory that the Sun formed the centre of the Solar System.

**FACT FILE**

- **1543**
  Copernicus publishes his book stating the theory that the Sun is the centre of the Solar System
- **1609**
  Galileo builds a telescope
- **1687**
  Newton publishes his book, *Principia*, containing the laws of motion and gravitation
- **1924**
  Hubble announces that galaxies exist outside the Milky Way

◀ *Newton explained that the force of gravity is responsible for keeping the planets and their moons in place*

# Famous astronomers

Sir Isaac Newton (1642-1727), an English scientist, made a revolutionary discovery while sitting under an apple tree. An apple fell down from the tree, making Newton wonder why it did not go up instead of falling down. This simple question led to one of the greatest discoveries ever – gravity. Edwin Hubble (1889-1953), an American astronomer, made another path-breaking discovery. He was the first to put forth the theory that the Universe is expanding. Stephen Hawking (1942- ) has made important discoveries about gravity and black holes. He is regarded as one of the greatest physicists of the 21st century.

# Satellites and Probes

**Man strives to learn more and more about space and everything in it. Apart from looking up at the sky and watching it through telescopes, we also send satellites and space probes to bring back information to Earth.**

## Eyes in the sky

Satellites are objects that orbit a planet. The moon is a natural satellite. We also send satellites into space. These satellites are called artificial, or man-made, satellites. Like the moon, man-made satellites too revolve around the Earth or any other planet. These help us to learn about the planets they orbit.

A probe is an unmanned spacecraft that collects information about objects, other than the Earth, in the Universe. Many satellites also come under this category.

*Satellites are of different kinds. Communication satellites are used to send radio and television signals. Weather satellites are used to predict weather conditions. Navigation satellites help ships and aircraft to find their way*

# Repairing a satellite

Satellites are machines and are prone to breakdowns. But how does one repair an object that is continuously circling the Earth? The only way to repair satellites is to go into space and fetch them. Now, this may seem like a hard task but space technology has made this possible. Space shuttles are sent to get these satellites back.

*Astronauts use a mechanical arm attached to the shuttle and grab hold of the satellite, locking it down on to the shuttle. The astronauts then make several space walks to repair or replace faulty systems*

# Runaway satellites

Satellites usually stay on course once they have been put into orbit. However, at times a satellite could stray from its orbit. In this case, the runaway satellite does not fall back on to Earth. Instead, the particles surrounding it along with the Sun's energy will cause the satellite to burn up. If it goes off course, in a direction opposite to the Sun, the satellite would continue to float in space, becoming a piece of space junk.

# Exploring Space

**For years, humans were satisfied merely looking up at the heavens and studying the Sun, the Moon and the stars. However, in the last 40 years or so, technology has developed by leaps and bounds, helping us to leave our planet and actually visit other objects in the Universe.**

## Where it all began

After the end of World War II in 1945, both the United States and the Soviet Union began space programmes to build rockets powerful enough to make space travel possible. A race to travel into space with unmanned probes and manned spacecraft began between the two countries.

▼ *The Russians were the first to send a human being to space. Cosmonaut Yuri Gagarin went into orbit on April 12, 1961, aboard Vostok 1*

## Blast off

On October 4, 1957, the Soviet Union became the first country to send an artificial satellite, called Sputnik I, into space. The United States sent their first satellite, Explorer 1, on January 31, 1958. Another American satellite, Pioneer 4, sent in March 1959, was a Moon probe that ended up in an orbit around the Sun and continues to be there.

*On May 5, 1961, Alan Shepard became the first American to enter space. He was on the Freedom 7 spacecraft*

*The first men to land on the Moon spent over 21 hours walking on its surface, exploring and collecting rocks to bring back to Earth*

## Men on the Moon

The Americans were the first to land on the Moon. On July 20, 1969, two American astronauts, Neil A. Armstrong and Edwin E. Aldrin, set foot on the Moon for the first time. Their vehicle was Apollo 11. Michael Collins, the third astronaut, continued in orbit in order to conduct tests and take photographs.

# Shuttling into Space

Flying into space is not as easy as flying an aeroplane. It is much more complicated and requires a lot of speed and skill. Astronomers use rockets to propel space shuttles into space.

The first-ever liquid-fuel rocket was built by Robert Goddard (1882-1945), an American scientist. It was launched for the first time on March 16, 1926

## Up, up and away

A space shuttle looks like an aeroplane. It is used to launch satellites into space. It also brings back damaged satellites and carries people and supplies to space stations. A space shuttle comprises an orbiter, two rocket boosters, an external fuel tank and two smaller fuel tanks. Only the orbiter and the smaller fuel tanks go into orbit. The rockets fall off shortly after lift-off. Once in space, the external fuel tank too is discarded. Space shuttles can land like aeroplanes on a runway.

The orbiter goes into circular orbit

External fuel tank falls off at 130 kilometres

Rocket boosters released at an altitude of 4 kilometres

## Naming a shuttle

NASA has three active space shuttle programmes – Discovery, Atlantis and Endeavour. Discovery is named after two famous ships of the same name. Atlantis is named after a research vessel, while Endeavour takes its name after the first ship commanded by James Cook, who is famous for his discovery of New Zealand, survey of Australia, and navigation around the Great Barrier Reef. In addition, NASA also owns the Enterprise and Pathfinder space shuttles.

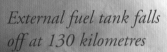

# Breaking the barrier

One of the biggest challenges that astronauts have to face is Earth's gravity. The gravitational pull is so strong that only a speed as high as 40,000 kilometres/hour (25,000 miles/hour) can launch a shuttle into space. A mixture of hydrogen and oxygen fuel, burned under high pressure, helps rockets achieve this speed. Once the rockets have pushed the shuttle into space, they drop back on to Earth. They are released into the sea, from where they are recovered to be used again.

## INTERESTING FACT

Space shuttles circle the Earth at speeds of about 28,164 km/h (17,500 mph). So it takes a shuttle only 90 minutes to go around the whole world. Moreover, the crew of a space shuttle can see a sunset or a sunrise every 45 minutes!

## FACT FILE

- **Heat of a shuttle during flight** 816°C, more than seven times the temperature of boiling water
- **Launch weight of the shuttle** About 2.04 million kg (4.5 million pounds)
- **Number of crewmembers** Usually 5-7, with capacity for 10
- **Maximum height a shuttle orbits at** About 965 km
- **Orbiter can be as long as** Over 37 metres (122 feet)
- **Height of the orbiter** About 17 metres (56 feet)

External fuel tank

Rocket booster

Orbiter

▼ *Solid fuel burns for two minutes*

# Space Fashion

We always wear clothes and shoes that are best suited to the weather outside. During summer we wear light clothes that keep us cool. In winter we take out all our warm woollens. A journey to space also requires special suits.

▲ *The spacesuit supplies astronauts with oxygen and guards them from temperature changes in space. The suit also protects against small meteorites*

## Clothes for all missions

Astronauts have more than one suit for space travel. When the spacecraft exits or re-enters the Earth's atmosphere, astronauts wear a special suit with a parachute pack. This suit has a helmet, gloves and boots. Inside the shuttle, astronauts wear comfortable clothes like knit shirts, pants or flight suits. They also carry special jackets, sleep shorts, slippers and undergarments, all of which have a special lining to protect the wearer.

▲ *Astronauts use a special backpack filled with nitrogen that helps them fly about freely in space. This pack, called the 'manned manoeuvring unit' (or MMU), is attached to the spacesuit. There is also a camera in it so that the astronaut can take pictures while in flight*

# Outside the space shuttle

While working outside the shuttle, astronauts wear a special suit called 'extravehicular mobility unit', or EMU. This suit is designed to be flexible and has several parts. Each part is made in various sizes and can be assembled to fit any astronaut. The undersuit contains pipes that allow water to flow over the body. This system keeps the astronauts cool. Astronauts also wear dark, mirrored visors to shield their eyes from the Sun's glare.

# Suited for space

The conditions in space are obviously very different from those we face on Earth. Hence, spacesuits have several features that protect astronauts from the harsh conditions. These special space dresses are tough and made of fabrics like nylon and Kevlar, a kind of rubber. Spacesuits are very heavy on Earth, but due to the absence of the effect of gravity in space, they become weightless.

# Living in Space

None of our normal activities can be carried out easily in space. This is because gravity, the force that keeps our feet firmly fixed on Earth, is almost completely absent in space. Can you imagine brushing your teeth, eating, or going to the bathroom in space? Astronauts have a tough time adapting to the special conditions in space.

## Keeping fit

Lower gravity in space can be quite unsettling. It causes several complications in the human body, including motion sickness, loss of bones and decreased rate of normal bone formation. Hence, astronauts are required to exercise regularly. They do so using devices such as a treadmill, a stationary bicycle, or a rowing machine.

◄ *It is vital for astronauts to remain fit during their space sojourns, since the lack of gravity effects cause various physical disorders*

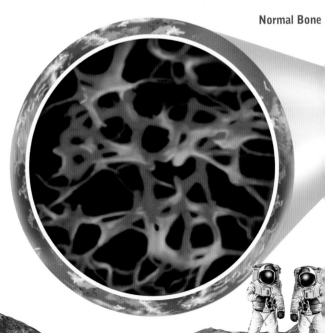

**Normal Bone**

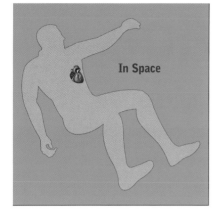

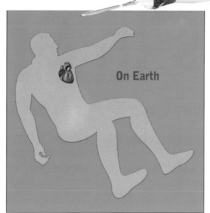

# Everyday living

Inside a spacecraft, even the food tends to float away. Usually the water in the food is removed so that it may weigh less and is easy to store. Sleeping in a space shuttle can be fun! You can float around and bounce against walls now and then. However, most astronauts prefer to be zipped into a sleeping bag. A little pillow is strapped to their heads.

**Osteoporotic bone**

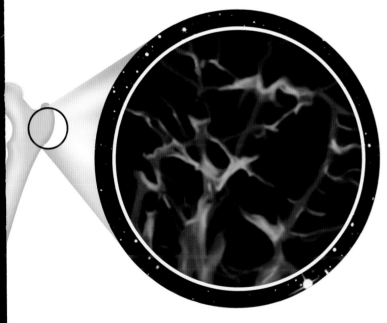

▲ *In the weightless environment of space, astronauts suffer from bone loss. This is a condition similar to a disorder called osteoporosis*

# Cleaning up

Diseases can spread easily among crewmembers in space. Hence, the living quarters and toilets are cleaned regularly. Dirty clothes and trash are stored in vacuum-sealed plastic bags and brought back to Earth. The toilet carries waste away from the astronaut's body into containers that are then sealed. Astronauts usually clean themselves with sponge baths.

# Accidents in Space

**In the past century, humans have made great strides in the field of space exploration. But success has come at the cost of the lives of many brave astronauts as well as ground crew. Going into space can be as dangerous as it is exciting.**

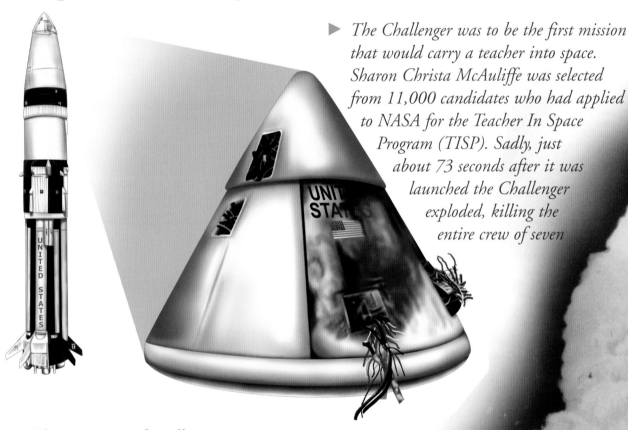

▶ *The Challenger was to be the first mission that would carry a teacher into space. Sharon Christa McAuliffe was selected from 11,000 candidates who had applied to NASA for the Teacher In Space Program (TISP). Sadly, just about 73 seconds after it was launched the Challenger exploded, killing the entire crew of seven*

▲ *The astronauts of Apollo 1 were trapped inside when a fire broke out*

## First failures

One of the first space missions that ended in an accident involved the Apollo/Saturn 204, later known as Apollo 1. The crew were on a training exercise when a fire broke out, killing all three. The first in-flight accident was that of the Russian spacecraft Soyuz 1. During re-entry the capsule's parachute failed to open and it crashed in a field, killing its single cosmonaut, Vladimir Komarov.

# Tragic end

Another tragic in-flight accident involved the Russian spacecraft Soyuz 11. In June 1971, the crew had become the first to dock at a space station, Salyut 1. After a three-week stay on the space station, the crew un-docked to return home. However, a valve on the spacecraft accidentally opened, draining the air in the craft out into space, killing all three on board.

*The death of the cosmonauts aboard the Soyuz 11 was discovered by the recovery team after the capsule had landed on Earth safely*

## Columbia explosion

On the day the space shuttle Columbia was expected to return home, the crew woke up to the tune of *Scotland the Brave* played in honour of mission specialist Laurel Clark's Scottish roots. The astronauts had spent 16 days in space. As Columbia re-entered the Earth's atmosphere, it exploded, killing all seven members on board.

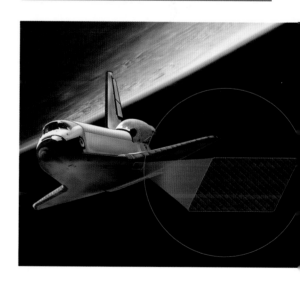

*The high-temperature surface insulation tiles cover the underside of the shuttle and can protect it from temperatures up to 2,300 degrees Fahrenheit* ▲

# Space Snippets

**Here are some interesting facts on space, the final frontier**

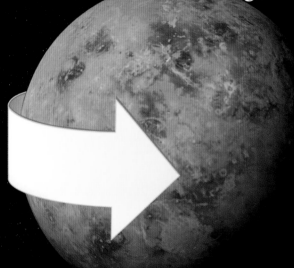

- The planet Venus spins in the opposite direction to other planets in the Solar System.

- The brightest star in the sky is Sirius. Also known as the Dog Star, its distance from the Solar System is about 51 trillion miles, or 8.6 light-years.

- To reach outer space, you need to travel at least 80 kilometres from the Earth's surface.

- The planet Mars was named after the Roman god of war. The month of March is also named after him.

- For centuries, people thought the appearance of a comet was an evil sign that could foretell the onset of plagues, wars and death.

- The largest asteroid on record is Ceres. It is so big it stretched over 965 kilometres!

- The planet Jupiter has no solid surface, only layers of gaseous clouds. It is composed mainly of hydrogen and helium.

- The Earth spins faster on its axis in September than it does in March!

- On March 29,1974, Mariner 10 was the first spacecraft to fly by the planet Mercury. It sent back the first close-up pictures of the planet's surface.

- The surface temperature of Venus is hot enough to melt lead! The surface can reach a temperature of 462°C. Lead melts at 328°C.

- Over 99 per cent of our Solar System's mass is concentrated in the Sun.

- Gemini 3, launched on March 23, 1965, was America's first two-man space flight. On that flight, Virgil 'Gus' Grissom became the first man to fly in space twice.

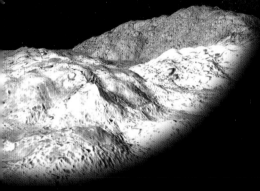

The Moon has about 3 trillion craters that are larger than 3 feet in diameter!

It takes Jupiter nearly 12 Earth years to orbit the Sun. The length of a day on Jupiter is close to 10 hours.

Yuri's Night is celebrated the world over on April 12 every year, in honour of Yuri Gagarin, the first man to go into space.

On January 16, 2003, Colonel Ilan Ramon became the first Israeli astronaut to go into space when he took his place aboard the ill-fated Columbia spacecraft.

- Golf is the only sport to have been played on the Moon. Astronaut Alan Shepard hit a golf ball on February 6, 1971, after landing on the Moon.

- If the Sun were to stop shining, people on Earth would only realise it eight minutes later.

- The phrase 'My Very Eager Mother Just Served Us Nine Pizzas' is used to remember the order of the nine planets, based on the first letter of each word – Mercury, Venus, Earth, Mars, Jupiter, Saturn, Uranus, Neptune and Pluto!

- The United States sent two mice, Benjy and Laska, into space in 1958.

- The main engine of a space shuttle may weigh only as much as a seventh of a train engine, but can produce as much power as 39 locomotives.

- On January 31, 1961, three-year-old Ham became the first chimpanzee to go into space.

- Marion Moon was the name of the mother of Edwin 'Buzz' Aldrin, the second man to set foot on the Moon.

# Glossary

**Abundant:** Existing in plentiful quantity; occurrence in relation to other elements or minerals in a specific environment

**Astronaut:** A person trained to travel and work in a spacecraft

**Axis:** An imaginary line about which an object rotates

**Capsule:** Part of a spacecraft that is used to transport humans or animals in outer space; vehicle designed to collect data from space and celestial bodies, and to be retrieved on returning to Earth

**Collision:** A brief event in which two or more bodies or particles come together, accompanied by an exchange of energy and sudden change of direction

**Contract:** To make or become less or narrower

**Crater:** A large bowl-shaped depression in a surface made by the impact of a body such as a meteoroid

**Dense:** Having high density – i.e., the amount of matter per unit volume

**Dock:** To connect two or more spacecraft in space

**Electron:** Matter with negative charge found in all atoms and orbiting the nucleus

**Elliptical:** An oval shape, resembling a flattened circle

**Equator:** An imaginary circle around the Earth placed at equal distance from the North and South poles, and dividing the North and South hemispheres

**Expand:** To increase in size, volume, quantity or scope

**Gravity:** The force of attraction that pulls objects in space towards each other, or towards the centre of a celestial body (the Earth, for example)

**Launch:** To set a spacecraft into motion; to propel with force

**Light-year:** The distance travelled by light in one year, calculated to be about 10 trillion kilometres

**Manned:** Spacecraft with a human crew

**Mesosphere:** Part of the upper atmosphere, from about 50 to 80 kilometres above the Earth's surface

**Motion sickness:** A state of feeling dizzy or nauseous caused while travelling in a moving vehicle

**NASA:** National Aeronautics and Space Administration, an

44

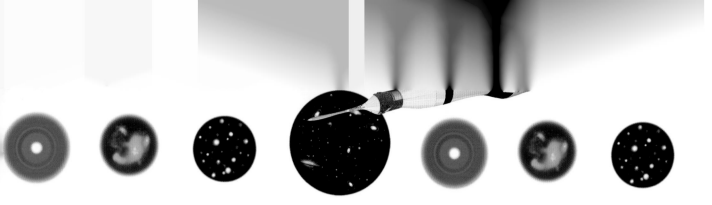

ndependent agency of the United States government for space research and travel

**Nebula:** An interstellar cloud of gas and dust that is visible either as a hazy patch or as a dark region

**Orbit:** The curved path of a celestial body or an artificial satellite as it moves around another body under the force of gravitation

**Osteoporosis**: A disorder marked by the thinning of bones, making them weak and prone to be easily broken

**Proton:** Matter with positive charge found in the centre of an atom

**Protostar:** A cloud of interstellar gas and dust that gradually collapses to gather into a hot dense core and eventually evolves into a star

**Re-entry:** The return of a spacecraft or a missile into Earth's atmosphere

**Revolution:** A complete orbital movement of a body, such as a planet or a satellite, around another

**Rotation:** The process of spinning about an internal axis

**Satellite**: A celestial body orbiting around a planet or star; a man-made object orbiting around the Earth or another celestial body, used for obtaining scientific data or for communication

**Space junk:** Man-made objects from spaceships and satellites that remain in Earth's orbit and serve no useful purpose

**Space walk:** To float and perform an activity outside a spacecraft in space, while being attached to the vehicle

**Stratosphere:** Upper region of the atmosphere ranging from a lower boundary of about 6-17 kilometres to its higher boundary at about 50 kilometres

**Thermosphere:** Region above the mesosphere, touching an altitude of about 400 kilometres and marked by increasing temperature

**Troposphere:** Lowest part of the Earth's atmosphere going up to about 10-13 kilometres above the surface

**Vacuum-sealed:** Packed in an airtight container under low pressure for preserving freshness

**Zodiac:** Imaginary area in the sky in which the Sun, the Moon and the planets appear to lie, and which has been divided into 12 equal parts. These parts are called signs of the zodiac, each with a special name and symbol